ICE WORKS CONSTRUCTION GUIDES

Falsework

C. J. Wilshere

ALUMNI/LYDON LIBRARY UNIVERSITY OF LOWELL

Thomas Telford Limited,
London, 1983

CONTENTS

Other ICE Works Construction Guides available
The Resident Engineer, J. Ballantyne
Pile driving, W.A. Dawson
Earthworks, P.C. Horner
Trade unions in construction, Will Howie
Site handling equipment, J.R. Illingworth
Tunnelling, T.M. Megaw
Site investigation, A.D. Robb
Concrete: materials technology, J.F. Troy
Access scaffolding, C.J. Wilshere

TH 5281 .W54 1983
Wilshere, C. J. 1925-
Falsework

Published by Thomas Telford Ltd, Telford House, PO Box 101, 26–34 Old Street, London EC1P 1JH

First published 1983

British Library Cataloguing in Publication Data
Wilshere, C.J.
Falsework (ICE works construction series)
1. Scaffolding 2. Shoring and underpinning
I. Title II. Series
624.1'6 TH5281
ISBN: 0 7277 0150 9

© C.J. Wilshere, 1983

All rights, including translation, reserved. Except for fair copying, no part of this publication may be reproduced, stored in a retrieval system or transmitted in any form or by any means electronic, mechanical, photocopying, recording or otherwise, without the prior written permission of the Managing Editor, Publications Division, Thomas Telford Ltd, PO Box 101, 26–34 Old Street, London EC1P 1JH

Typeset in Great Britain by MHL Typesetting Limited, Coventry
Printed and bound in Great Britain by H. Charlesworth & Co. Ltd, Huddersfield

1. INTRODUCTION

Falsework is the temporary structure which enables the permanent structure to be constructed, and which must be retained until the permanent structure is self-supporting.

Some form of falsework has been in use ever since mankind started to build. The Romans constructed a great many arches and domes and there has been much speculation as to exactly what form their falsework took. Similarly, medieval cathedrals and other buildings involved falsework of some complexity. The common theme here is arch construction; an arch requires full support until the keystone is in position. Throughout history construction of any significance was carried out in stone; this type of full support, or centering as it is often known (Fig. 1, page 3), was representative of most falsework. The advent of concrete construction has broadened the scope of falsework enormously, both for in situ and precast construction, and many modern structures of timber, steel and brickwork need temporary support during their construction.

This booklet covers falsework of all magnitudes, from placing the lintel in a house wall to the construction of a major bridge. The wide range of material used in falsework is discussed including proprietary equipment as well as timber, steel and aluminium. The definition of falsework given above is wide-ranging but it is not normally applied to support works for excavations such as coffer-dams and trenches. Nor is it applied to formwork and it is difficult to find a universally acceptable definition which explains the division of formwork from falsework. Concrete has to be contained until it hardens, and the facing surface of the container is called formwork. Use of the word formwork is usually extended to include the immediate supports and in casting walls for example, the whole structure would be considered to be formwork, though many of the points discussed in this Guide would apply.

Falsework primarily presents a technical problem. However, experience has shown that failures of falsework take place for many reasons and a significant cause of failure is that the logical and accepted steps or stages in the design and construction of falsework are not correctly followed. For this reason it is essential to have administrative processes to help ensure that the correct sequence of activities and appropriate checking of the actual falsework are carried out. Section 8 discusses procedures and ways in which this subject can be tackled.

As this is an introductory guide to the subject, it provides guidance on other sources of information, without discussing the more complex and unusual approaches to falsework. Some guidance and advice is given on the more commonly used equipment and methods, but it is not a design guide.

2. REQUIREMENTS

Falsework has to meet three main requirements. It should provide a safe support, that is to say there should be neither a failure nor risk of one. The second requirement is that it should provide this support without undue movement, that is to say the permanent structure should be within the tolerances laid down and without locked-in stresses. Thirdly, the cost should be as small as possible. In many ways the first two are contradictory to the third and satisfactory falsework has to be a reasonable compromise between these opposed requirements.

The structure being built, referred to as the payload, must be supported by the falsework. This will normally be a static load and the assessment of its mass should give little difficulty. In addition there will be other vertical loads.

Operatives, supervisors and other people will create a mass which must be added to the loading of the falsework. Similarly, there will be tools small or large; there may be other ancillary loads such as concrete pipelines; and there will be concrete heaped prior to its being spread. In extreme cases there may be dumpers moving on the falsework. For design purposes a figure of 2 kN/m^2 is normally adopted to cover a traditional labour force and concreting methods. Other equipment (pumps, dumpers) should be assessed specifically and an allowance made where there is any movement which will cause horizontal as well as vertical loads.

Horizontal loads must include wind loading. It may be assumed that construction will continue until the wind has reached a Beaufort wind speed of six. This is described as 'whole branches of trees in motion', and it is unreasonable to expect people to remain on a site where things can be blown about. If it is part of the design to adopt such a figure, then a procedure must be adopted to ensure that operations stop at this intensity of wind. It will normally be necessary to stop the operations of any cranes at this velocity. However, the falsework itself will remain in whatever gale may arise and an appropriate design force has to be taken. This will range from as little as 400 N/m^2 in a very sheltered corner of London to 1350 N/m^2, or perhaps more, in the windiest parts of Scotland, depending on locality, height and exposure. The *Code of practice for falsework*, BS 5975,[1] enables the appropriate calculations to be carried out for most circumstances.

Other horizontal loads which should be considered are the constructional inaccuracy load of 1% minimum (it should be taken as $2\frac{1}{2}$% when there are as few as four vertical members in the falsework); loads from water, perhaps from waves; and impact loads from passing plant, vehicles, and so on. In large structures it is necessary to consider the problems of thermal movement and shrinkage of the concrete. Fig. 2 (page 4) gives an idea of the various types of load which arise.

It is not sufficient to consider these loads only at the final loaded state. There may be various stages during the process of construction which are more critical for one part of the falsework or another and, similarly, there may be problems during dismantling. It is essential to consider all the stages, and if there are critical ones, the instructions to the people on site must include guidance on the sequence and any particular precautions which may have to be taken.

Access must be safe and conducive to efficient working. It is all too easy to conceive and design falsework without reference to the people who have to construct it, but provision should always be made for getting people from the

Fig. 1. Centering to Blackfriars Bridge 1766

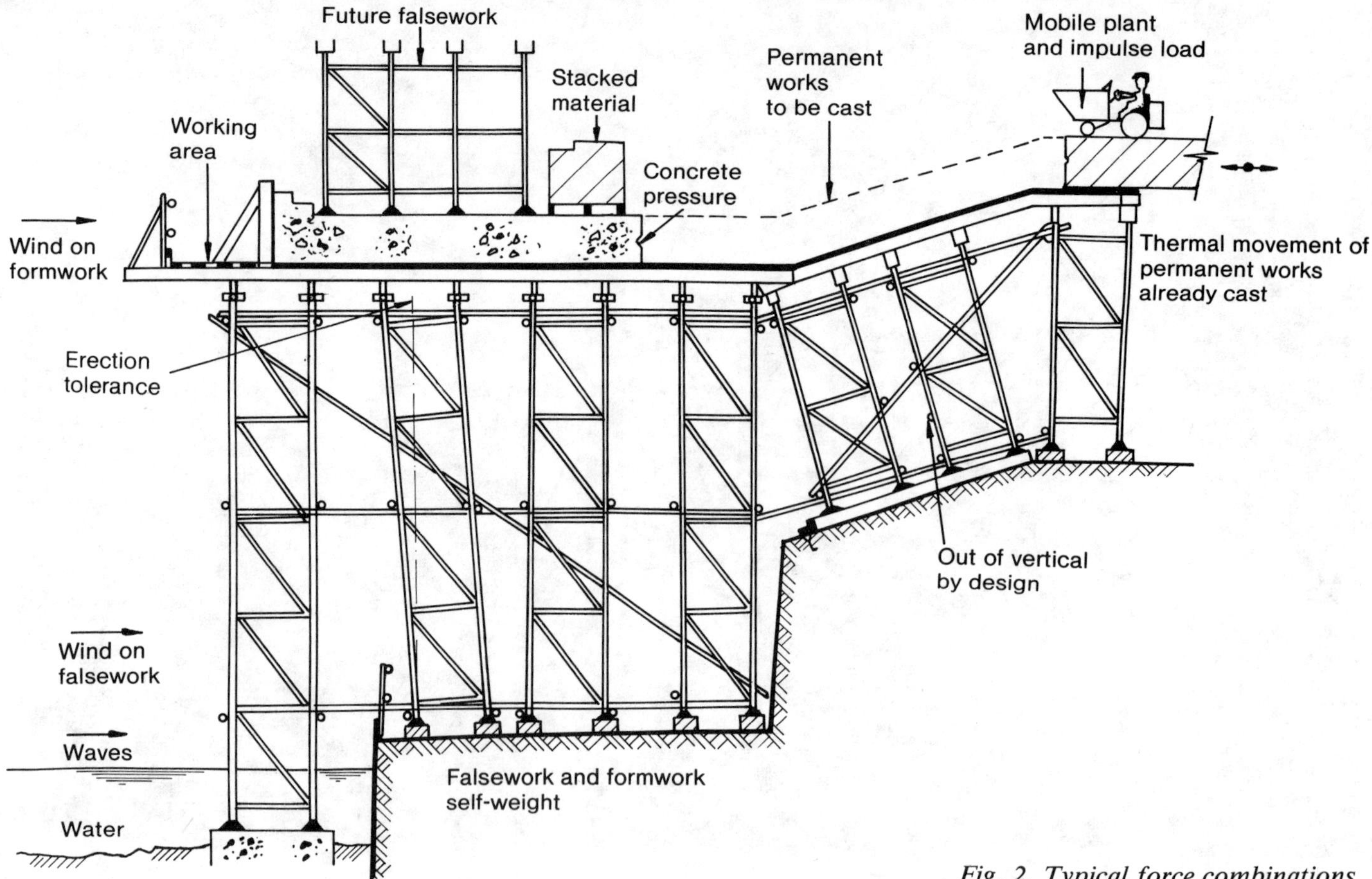

Fig. 2. Typical force combinations

ground to where they are going to work; the platform must be hazard-free and at a suitable level for the particular activity to be serviced from it.

3. MATERIALS AND EQUIPMENT

The materials used for falsework may be fabricated or prepared on- or off-site. The choice between site fabrication and manufacture elsewhere reflects a number of points. On site, the operation is directly under the control of those who are going to use the falsework and there should be a more realistic appreciation of the urgency; in many cases a greater degree of adaptation is possible. However, what it is possible to do with site facilities is limited, because good workshop space is at a premium as is the availability of skill and equipment. The site can obtain ready-made items, which are usually proprietary, and for which all the design and fabrication have already been done. This enables those on site to concentrate on aspects of the job which they cannot choose to have done elsewhere. Off-site facilities offer a wider range of possibilities for purpose-made items. It may be necessary to inspect what is being made at the works.

Timber

Wood or timber has been in use longer than anything else. It must be of adequate quality and in adequate condition. For many years, reliance was placed on experienced joiners to eliminate the worst pieces of timber from positions where they would be significantly loaded, and, with traditional apprenticeships and experience, this proved to be a fairly satisfactory arrangement. Nowadays such skill is found infrequently, so it is seldom possible to rely on this sifting process. It may well fall to the engineer to check that the timber is of the quality assumed in the design and to sort through to make sure that the risks of the substandard piece being used are minimised.

The Code of Practice on timber, BS 5268, Part 2,[2] currently CP 112,[3] provides information on design stresses and modification factors. Stresses are modified to accord with site conditions; working figures so calculated are given in the Falsework Code.[1] The use of timber on site varies from the long-term structural application in three main respects. Firstly, it will be loaded for a very much shorter time, and provided there is a time lapse between individual loadings, the effect of these loadings is not cumulative. Secondly, timber on site is 'wet' because its moisture content is invariably over 20%. The third point is connected solely with bearing. BS 5268[2] accepts that a proportion of timber will be waney, which means that some of the corners will be missing where the bark was. Because this has a significant effect on the bearing capacity, it is a normal requirement in falsework design to exclude the use of waney timbers in bearing.

The Timber Code,[3] concerns itself only with stress-graded timber, graded either visually or by machine. There are three grades (and a reject fraction) for each species, and permissible stresses are given for each of these combinations. Where a particular species and grade is adhered to completely, no problems should arise as to which timber is being used. But if different timbers are acquired at different times there can be problems. In practice, the fresh delivery gets mixed with last month's delivery, and in turn this will all get mixed up with next month's delivery. As soon as the markings can no longer be read, all timber will have to be considered the same as the weakest timber known to have been delivered.

The revised Timber Code[2] offers a simpler approach. Instead of this multitude of strength classes created by dealing with each timber species separately, a small number of strength levels are proposed and each piece of timber is then classified into one or other of these. Thus, timber of different species could be obtained from a variety of sources, but could be put in the same pool for general application. This approach is put forward in parallel with the existing stress-grading approach.

The concept has been adopted in the Falsework Code[1] and the three relevant classes are SC3, SC4, and SC5, in ascending order of strength. Appropriate permissible stresses are given and the lowest is taken as the general typical level.

Unless timber is acquired at a particular grade (and the markings are still visible), it cannot be assumed to be all as good as SC3. It is necessary to establish that such timber has the required strength. The rules of BS 4978[4] for visual stress grading can be applied by people who have the right training, but it is not an easy task for the novice. TRADA have produced a leaflet[5] which sets down simpler rules which apply to the commoner timbers and which will enable an engineer to identify and reject timber worse than SC3. Identification is made by comparing the knots, wane, slope of grain, rate of growth, fissures and other features with the acceptable levels set out. But the rejection rate for mutilated timber is too low even when this method is used. Check against the following before re-using timber.

(*a*) The timber should comply with the specification both for quality and dimensions. Unless specifically ordered, two pieces of timber may have slightly different cross-sections, sufficient to vitiate any design assuming them to be the same. This applies whether the timber is rough-sawn or planed.
(*b*) The timber should not be painted in such a way that the state of the timber is obscured.
(*c*) Rot, which is shown up because it is softer than the surrounding timber, should be rejected. It may appear as a white or brown discoloration.
(*d*) Any cut in the edge greater than 1/20th of the thickness of the section is unacceptable.
(*e*) Holes in the outer two thirds are also unacceptable.
(*f*) Check any other mechanical damage and reject the timber if in doubt.
(*g*) Where a piece of timber is to be used as a beam or a column it should not have grain at slopes greater than 1 in 6 for class SC3 or 1 in 10 for class SC4. A swivel-scribe is needed to check this as the surface appearance does not give the direction of the grain.

Rough-sawn timber should never be more than 1 mm undersize, and planed timber should never be more than 6 mm less than nominal size.

Steel

Steel is commonly used on site as rolled sections, such as joists, universal beams and columns, or structural hollow sections. It is familiar as scaffold tube and its allied fittings, and in general, as used in workshop fabrication, it can be obtained in a number of sections. It also varies in grade. The basic grade used today for rolled sections is 43 and it is still often referred to as mild steel. In practice this is almost always 43A. It is also practical to get rolled steel sections in grade 50 and sometimes in higher grades. There are many special steels and some of very much higher strength are available. However, where one property such as strength is particularly good, another will be poorer; thus, for example, for high strength steel the welding rules are more stringent. Grade 43 is the most-used compromise between the various conflicting possibilities.

Any welding needed should be done to suit the appropriate grade of steel. When there is any doubt as to what the steel is, it is normal to specify that welding is suitable for a grade 50 steel, while design is carried out to grade 43 unless it is certain that the material on site is of greater strength. Unfortunately, there is no simple means of knowing the grade of steel and if it is essential to know, a sample must be sent to a laboratory. In practice, the design of steelwork for temporary works is often controlled by the deflection. Because the modulus of elasticity of all steel is the same, there is no advantage in buying a stronger steel where deflection is the controlling factor. It will also be apparent that the risk of substitution is significant, and that where it is intended to use a higher grade steel, particular attention should be paid to ensuring that it is actually being used. Inadvertent mixing of material can also happen because different rollings of the same basic size have different thicknesses of web and flange, and thus different strengths. With the aid of a section book and a caliper gauge to measure the thickness at the end of member, it is possible to identify positively any particular section.

Steel must be inspected before use to ensure that it is to specification and in proper condition. Even steel delivered directly to the site in 'as new' condition can be significantly distorted, but this is particularly true of steel being reused.

Limits of accuracy

Straightness. This is measured as the offset from a line joining the ends.

For all members:	where $L < 3350$ mm, $\Delta < 5$ mm
For beams:	where 3350 mm $< L <$ 26,666 mm, $\Delta < 0.0015L$
	where $L >$ 26,666 mm, $\Delta < 40$ mm
For columns:	where 3350 mm $< L <$ 16,666 mm, $\Delta < 0.0015L$
	where $L >$ 16,666 mm, $\Delta < 25$ mm

where L is the length in mm and Δ is the offset. The total of the two angles between the actual and the true direction of the flanges should not exceed 1°.

Steel should not be accepted if it is significantly rusty; it can be found in this state when it is delivered 'new' from a

Fig. 3. Steelwork with web stiffeners

stockist. If there is any significant sign of loss of section, either the steel should be discarded, or the stresses used in design be reduced significantly. Damage must also be examined. Any cuts in the flanges or holes should be deducted from the working area and the effect of eccentricity considered. If the design has assumed complete sections, it will probably be necessary to discard the steel.

Rectification should only be undertaken by those who have adequate training and experience. The welding which may be attempted should be carried out in accordance with BS 5135.[6] Welds always need careful checking both when new to check quality and size and later to ensure they have not cracked.

Where steelwork is fully fabricated and welded or bolted together, the normal considerations apply. Much steelwork for falsework is less securely fixed than steelwork for permanent structures and is relatively short in length.

Lateral stability is much more of a problem: some lateral restraint should be provided unless the assembly is clearly stable against overturning. Because of this incipient lateral instability, the capacity of beams and cantilevers is reduced.

Web-buckling failures are possible particularly with shorter members and it is sensible to have web stiffeners welded in unless calculations clearly show them to be unnecessary (Fig. 3).

When two pieces of steelwork with flat faces are placed one on another, it is seldom they will mate perfectly. Some rotation can be expected when the falsework is loaded. Unless steps are taken, the load becomes concentrated along the edge. Web stiffeners may reduce the risk of failure sufficiently, but it is better to provide a detail which will give a correct load path.

One approach is to use a shallow bed of mortar where mating steelwork faces do not marry up because of twists or because a small intentional angular deviation occurs and movement is not expected. The mortar should be carefully cured; this can be done by putting it in a large plastic bag, and making sure that it overhangs the edges so making a key with the members. However, this is only appropriate where lateral loading is negligible. Where hinging is required, a narrow strip of steel can be tack-welded to thc centre of the lower joist, ensuring a concentric load path.

Scaffolding

Steel scaffold tube is normally made from a weaker, more ductile grade of steel than grade 43. British tube is 48.3 mm outside diameter and 4 mm thick. Much has very elementary rust protection, but there is an increasing amount which is galvanised. Aluminium alloy tube is available, with the same external diameter, but 4.5 mm thick. Largely because of its low modulus of elasticity, aluminium is seldom used for falsework and should never be mixed indiscriminately with it. It is useful in clean conditions. Both are specified in BS 1139 Part 1.[7] (See Fig. 4.)

Outside the British Isles, tubes of other thicknesses and other steel grades are in use as well. These have different carrying capacities and the information given in the Falsework Code[1] does not apply to them.

Scaffold tube is an economic material because it can be

Fig. 4. Tubular scaffold falsework for a water tower

reused extensively. To avoid cutting to length, which would be needed if all joints were in line, at most intersections tubes pass each other giving a small eccentricity. A variety of couplers are available to enable all the joints to be made simply and to make structurally sound falsework.

Fittings are made of steel, although it is possible to obtain some made of aluminium. Steel fittings can be used with aluminium tube.

A number of proprietary components are made with tube not of 48.3 mm diameter and it is possible to obtain couplers appropriate for a variety of sizes. Tube 63 mm nominal diameter is also available with its own family of fittings. Scaffold tube was originally used for access and it is discussed in the parallel Works Construction Guide, *Access scaffolding*.[8] There are some additional components.

Adjustable baseplate (Fig. 5, opposite)
A threaded stalk fixed to a square flat plate has a 'nut' with a shoulder, so that when it is inserted into a tube, the load is transferred through this shoulder. There is only a limited amount of thread, so that there will always be a portion inserted well into the tube, which gives some stability. It can be used inverted as a post-head.

Adjustable forkhead
This is very similar to the adjustable baseplate, but the plate is replaced by a U section. This enables beam members—timber or steel—to be retained, but the sides have very little strength, and should be regarded purely as a convenience and safety aid during erection. With both components, the capacity is frequently limited when they are near their full extension and there are very few which will carry their full load unbraced when extended more than 300 mm. For this reason there is often a short section of the stalk without any thread adjacent to the plate to which bracing may be fixed. The attachment point may be of normal scaffold tube diameter permitting standard fittings to be used, or it may be desirable to obtain special fittings. Consult the manufacturer's instructions. The size of the plate or fork may limit the load, because the timber bearing pressure would otherwise be too high.

The joint made between an adjustable unit and the tube which supports it permits a little angular movement. As these joints are all at the same height, this movement constitutes a weak point in the falsework and for this reason it is desirable that there is diagonal bracing connecting the top of the tubes to the top of the adjustable head.

Tubing
Tube to be erected should be in good condition. It should be straight; if rolled over when supported near the ends it should do so smoothly without any apparent deformity. It should have no creases or cracks of any sort. Where it has been bent, the bent section should be cut out and discarded if there is any sign of a crease. Otherwise it is possible to put it through a straightening machine, which will return it to within acceptable straightness. It should be reasonably free from rust and corrosion. As with structural steel, any significant pitting must result in the steel being downgraded or discarded. Deterioration of tubes

happens particularly over the 75 mm nearest each end, and if this is cut away, the full section of the tube will usually be seen again.

The fittings to be used in conjunction with scaffold tubes are specified in BS 1139, Part 2.[9] This gives a number of test loads, which each fitting should comply with, and it is normal to take a factor of safety of two with these test figures to give permissible working loads. These are listed in the Falsework Code.[1]

The inspection of fittings should ensure that they are

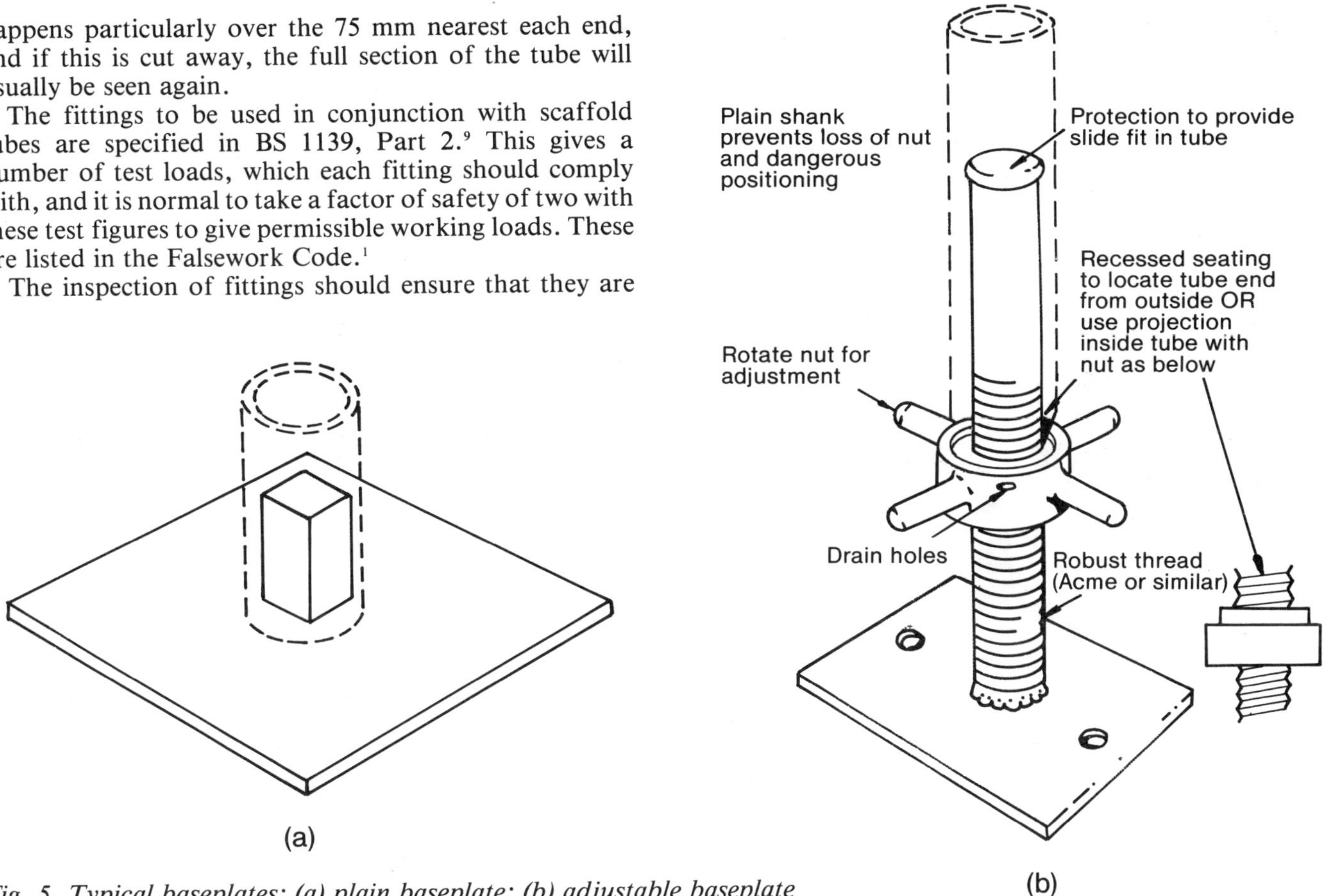

Fig. 5. Typical baseplates: (a) plain baseplate; (b) adjustable baseplate

visually undamaged and that any threads or moving parts work freely, so that there is no risk of tightness being assumed when in fact friction of threads is preventing a nut from moving. This applies to the threads of adjustables just as much, and they should be checked for trueness. Stiff threads can waste a lot of time.

While bracing for a scaffold framework may be connected with standard couplers, with its attendant joint eccentricity, it is essential that the load is brought on to tubes acting as struts as concentrically as practical. This is normally done with an adjustable forkhead and it may be necessary to have an adjustable base as well. Note that the load which is transmitted by a fitting in friction to a tube is normally limited to 6.25 kN, although a concentrically loaded tube can carry several times this. The Falsework Code[1] provides information.

Fig. 6. Aluminium beams in use for supporting a soffit

Aluminium

Despite the particular disadvantage of relatively high flexibility, aluminium beams are now in use (Fig. 6). The main attractions are their lightness and the fact that they offer an opportunity to design complex sections to make connections much simpler. They are produced by the extrusion process which permits complicated shapes to be made reasonably cheaply. It should be noted that there are several similar proprietary beams which could be confused. (See pages 9–10 for information on aluminium scaffold tube.)

Equipment must be in reasonable condition. Corrosion does not appear to be a significant problem, but experience of this material is relatively short and caution should be exercised. As with all structural sections there is the possibility of deformation and this possibility should be acknowledged. Bent beams should be straightened or rejected. Do not weld aluminium under site conditions.

Manufactured equipment

It is possible to design special equipment and have it made for a particular job. In practice over the years the items which originated in this way and which have proved to be generally useful have become available from manufac-

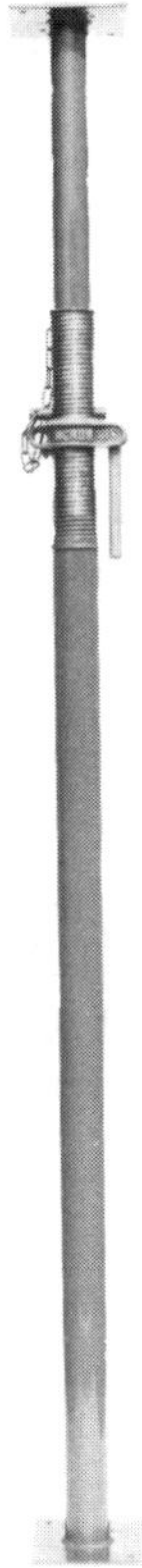

Fig. 7. An acrow prop

Range of lengths: m	Safe working loads: kN			
Overall extended lengths (props: No 0 1.07–1.82; No 1 1.75–3.12; No 2 2.00–3.35; No 3 2.59–3.96; No 4 3.20–4.87)	General (No 0–No 3)	Systems (No 0–No 3)	General (No 4)	Systems (No 4)
4.87			7	9
4.75			7	9
4.50			8	10
4.25			9	11
4.00			10	12
3.96	9	12	10	13
3.75	10	13	11	15
3.50	11	15	14	19
3.35	12	16	16	22
3.20	14	18	17	24*
3.12	14	18		
3.00	15	19		
2.75	17	23*		
2.59	18	25*		
2.25	18	32*		
2.00	18	35*		
1.82	18	35*		
1.75	18	35*		
1.07	18	35*		

*May be limited further by bearing stress

Fig. 8. Range of lengths and safe working loads of props

turers and form the basis of proprietary equipment as we know it today. The term 'proprietary' indicates that a design is unique to a particular manufacturer and carries his mark and warranty. It is possible to mix components from different manufacturers, but subtle differences could result in problems and the manufacturer's warranty would be invalidated. But there are some items which are made by a number of manufacturers which can be used together in practice and can no longer be called proprietary equipment. The best known of these is the telescopic prop, commonly called an acrow from the name of the firm which brought it to the United Kingdom in the 1930's (Fig. 7).

A prop consists of two tubes which telescope one within the other, each with an end plate of side 150 mm. The inner, the same diameter as scaffold tube, has a series of holes through one of which a special pin is inserted. The ends of the pin are supported on a collar which is screwed on to the end of the outer tube. A longitudinal slot permits fine adjustment. Coarse adjustment is obtained by putting the pin through a different hole in the inner tube. This prop is specified in BS 4074.[10] It is available in five sizes—see Fig. 8.

The strength of this type of prop was investigated by CIRIA[11] and safe working loads related to conditions of use are given in Fig. 8. A survey of site practice showed considerable deviation from the ideal standards of erection and the research indicated that there was a reduction in strength compared with the capacity of a prop erected perfectly. Props are erected with varying standards of plumb and the load is usually not brought on concentrically. By using a hand-level with a cross-bubble on a few props and eyeing in the rest, one can obtain the required standard. This is within 1.5° of vertical or 1 in 40. To ensure concentricity, a forkhead can be used, and twisted. Some soffit support equipment is designed to pin into the holes provided in the end plates of most props; this achieves the same result. Otherwise the capacities appropriate to eccentric loading should be used and care taken that the beams are within 25 mm of the centre. Fig. 8 shows the safe working loads for the two conditions.

It is usual to lace props over 2.75 m height, as a help in erection and to make the structure a little more stable. Unless this lacing is effectively located (in an engineering sense), there is no increase in prop capacity. Props should be unbent and have the correct high tensile pin.

Proprietary scaffold

A number of frame systems are available, which assemble into rectangular (or triangular) towers. The desired height is achieved by the choice of an appropriate set of frame heights, together with adjustable bases and head jacks. The vertical load-carrying capacity of each vertical member is assured by the cross-members which limit the effective length of the verticals, provided the tower is stabilised as a unit. For tall towers, this is a significant requirement. Loads of 6 t per leg are possible with certain makes. The British Standards Institution is drawing up a standard approach for testing such towers.[12]

Much use is also made of the 'pocket' scaffolds (Fig. 9). Here, each vertical has a number of positions for fixing the horizontal members, usually about half a metre apart. The connection consists of two mating components, one

Fig. 9. Prefabricated scaffolding as falsework

Fig. 10. Telescopic centres supporting plywood

..., trestling as falsework

being welded to each end of the horizontal member, and the other, the pocket, to the vertical. These connections are designed for quick operation and to give some moment capacity. Depending on the type of joint, the spacing of the horizontals, and the arrangement of the diagonals, the capacity per leg can be up to 5.5 t. Because the system is based on individual legs, towers or grids of any number of verticals can be set up, and, with the specified bracing, will be self-stabilising. However, this type of scaffold presents the biggest opportunity for mixing makes and the agreement of the manufacturers concerned should be obtained before adopting a mixed design.

It is important to follow the manufacturer's rules. Scaffolds of this type are difficult to analyse mathematically and full-scale testing is normally used to verify the data. It should be checked that the component being used is the one that was intended, that it is not bent and that the welds are still sound. Because some newer components are of metric length there is a possibility of confusion with the equivalent imperially dimensioned piece. As yet there is no agreed convention to distinguish them. The adjustable screwed components used with proprietary scaffold normally have a large extension, so the need to provide additional bracing should be assessed.

Telescopic centres

Because the spans of slabs vary, the use of fixed-length beam members is not very practical. Instead, telescopic beams or centres are available (Fig. 10). Spans vary from 1.2 m to 9.3 m, but at the largest spans telescopic beams may not be practicable because loads to be handled are too heavy. Normally, there are two units: an outer unit of rectangular or trapezoidal box-section, with sides made from a lattice of rods, and an inner unit which is slightly smaller so that it can fit inside. The inner unit may also be of lattice section or it may be a sheet-metal joist member. The top of these beams is solid to provide a support for plywood or panels. At the ends, lugs are provided level with the top for support. It may be practical to seat each lug in a pocket on the wall, or alternatively it may be necessary to erect a special support for it. The length of the lug is small, and care must be taken to see that it is well home on to its support. Where this is a timber bearer, it must be ensured that the timber cannot overturn, as the load is applied eccentrically. Bearing pressures on timber can be excessive with the larger centres. It is possible to increase the capacity by providing an intermediate support but because of the structure of the centre, this must only be done strictly in accordance with the rules for that particular model of centre. Information on this and on capacity should be obtained from the supplier. Reference 13 gives information on a standard test method to assess centres, with which a safety factor of 2.2 is traditionally adopted. Most centres have an inbuilt camber to offset the deflection under load and this can be adjusted by releasing the locking screw; this is a dangerous practice, because only a small amount of vibration can result in the centre moving far enough to drop off one of its supports, and it is much more preferable to lock it.

Because centres are not very rigid, it is difficult to have good quality construction joints in the middle of a concrete soffit supported by centres. This is because the first

Fig. 11. Bailey type bridging spanning a river

Fig. 12. Modular support girder with undertrussing

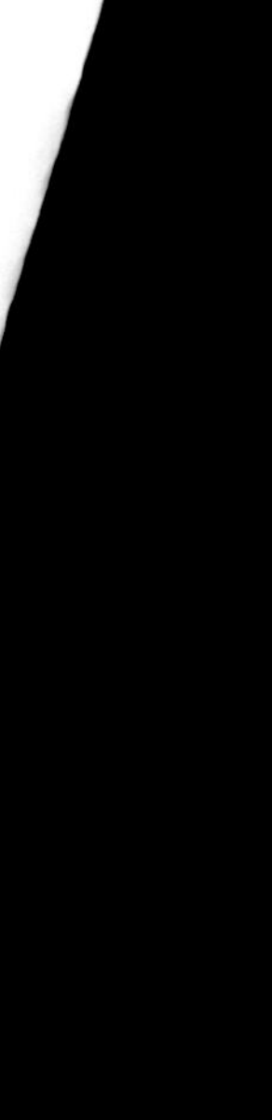

Fig. 13.

pour will cause partial deflection, and the second pour will make the centres deflect fully. The concrete already poured will not deflect, and some grout will run under the earlier concrete. However, centres have an advantage where access to the space below is needed. Similarly, it will be advantageous if temporary foundations are only available on or near the permanent supports. All these items are intended for man-handling, although the bigger ones will require some enthusiasm from those involved. For information on strength refer to the manufacturer's data.

Like other fabricated components, welds need to be examined; eventually rusting will become excessive. Bends, particularly likely in the lattice members, will render the centre unsafe.

Heavier duty horizontal members

The Bailey bridge (Fig. 11) has long been used and there are modern derivatives. (The *Bailey and Uniflote handbook*[14] deals with this subject.) Sagging normally occurs because of the methods of pinning members together and care has to be taken to obtain the shape required, taking into account both deflection and sagging.

There are other units available (see Fig. 12), designed specifically for temporary works, which are simpler to use. For all these, refer to the manufacturers for advice on design and use.

Heavy duty vertical members

Bailey bridging is also used to support large vertical loads. L, or military trestling (Fig. 13), is a further practical method which carries loads per leg of 40–80 t. It comprises vertical units of twin channels with end plates, which can be assembled to the desired height. Bracing members enable rectangular towers or larger assemblies to be made up. The accuracy of fabrication affects the capacity and some more recently made components effectively reduce the capacity to the bottom of the range.

There are also purpose-made items generally of tower configuration varying in capacity from 15 t to 160 t. All have adequate end components to transfer load efficiently. Most bolt together and so some tension loads can be accommodated. The manufacturers should be consulted for design proposals, information on inspection and assembly details. (See Fig. 14 overleaf.)

4. GROUND

While it is sometimes possible to use the permanent structure as a support for falsework, it is frequently necessary to use the ground. This may vary from solid rock (which will carry any load likely to be put upon it) to soft clay and mud or made ground where the capacity is very much smaller and much more uncertain.

The three main groups are rock, cohesive materials such as clay and granular materials such as sand or gravel. Rock usually provides the best support, but it may have cracks and cleavage planes which permit one piece to move on another. This situation is worse when there are

Fig. 14. Light duty Trishores supporting beams

clayey bands interleaved. Clay is subject to deterioration due to weather or water in the ground. Granular materials, which water permeates easily, become less stable when wet.

Before a design for falsework is carried out, an examination of the ground conditions is essential. This may be quite simple if an adequate soils report has already been prepared, but it is frequently necessary to conduct a survey particularly for the purpose. This examination should extend to one and a half times the width of the expected falsework structure and should be deep enough to ensure that there are no weak layers before the load has spread widely. (A spread of 45% can be assumed.) The type of ground should be determined; then it can be assessed for a safe capacity. Both these steps are dealt with in the Falsework Code,[1] in which table 16 (page 31) gives help with descriptions. A soil-mechanics specialist can then be consulted or the table in the Code used. The simplified method given in the Code consists of taking the appropriate figure for safe bearing pressure given in the table, and applying various factors: one for reliability of information, one for settlement and its consequences and a third for ground-water level. Where the water level will remain lower than the breadth of the foundation, it may be ignored. This will give a result which will quite often be conservative.

An alternative approach is to use an instrument such as a hand penetrometer. But even this requires some training to avoid pitfalls.

Naturally, attempts are made to reduce the amount of preparation to provide the necessary foundation. It is seldom that loads are light enough to be carried on topsoil, but even when this is stripped off for a third or half a metre, the ground underneath may not be adequate. On very soft ground, it may be wiser to leave matted vegetation to help distribute the load.

Fill material is very difficult to assess, because unless it is all dug up, there is no certainty that it is all the same. Thus it is frequently necessary to remove it.

Theory

Load must be spread laterally over the foundation so that the pressure at any level is acceptable. Where there is a high capacity surface like concrete or steel, no problem arises, but otherwise some means of load spreading is needed. Timber sole plates are often used, but unless they are stiff, that is to say thick, the lateral dispersion they give is small, because they bend. The ends of sole plates should be arranged as far as possible well away from the verticals. A conventional reinforced concrete base may be needed, or occasionally piles.

There are two main approaches to sloping ground. In the first, horizontal shelves are cut in the bank, and sole plates etc. put on them. This carries the risk of failure of the whole bank and expert advice should always be sought where the bank is steeper than 1 in 6.

Alternatively, a sole plate can be placed up the slope and the load from the verticals brought on to it by suitably shaped blocks. This method supports the normal component of the load and the tangential component is catered for either by struts on top of the sole plate, carried to an anchorage at the bottom of the slope, or with blocks

Fig. 15. Detail of scaffold falsework supported from a slope

secured to the sole plate to transfer the load in the same way (Fig. 15).

Preparation

The surface should be protected against the weather as soon as the excavation has reached a satisfactory level. This will be at the design depth if it will carry the required load, lower if the capacity is less than anticipated. A bigger spread may have to be adopted or a soft layer removed. Granular soils should be compacted. It is ideal to blind the surface with concrete and bed the sole plate down in the blinding. In all cases the sole plate should be in full contact with the ground, with no soft spots. A fill such as hoggin well compacted can be an alternative to blinding concrete.

The need to keep foundations in the condition in which they are found is important. They should not be exposed to the weather. It is most desirable that water be kept away from them, both to prevent softening as explained, and, where the water is flowing, to prevent erosion. Thought should always be given to the water course which will form naturally in a newly excavated site, should there be a cloud burst.

Particular points of risk include

(*a*) complex supports and supports in water courses
(*b*) where there is a weaker area below
(*c*) frozen ground
(*d*) sloping sites
(*e*) compaction of granular material due to vibration from piling and so on
(*f*) where different parts of the falsework are on different types of material

It is not practical to learn how to judge soils from a book and it is essential that those called upon to approve foundations for given capacities should have had some practical training in this subject.

Where the support is from another part of the structure, it is most important to check that it is strong enough. The overall capacity may well be adequate, but individual slabs are often too weak, and it may be necessary to share the load between two or more slabs. When part of the support is from the permanent structure and the remainder is from the ground, there is a possibility of differential settlement.

5. CONCEPTS AND DESIGN CONSIDERATIONS

There are frequently several solutions to a particular falsework problem, and a choice should be made before proceeding with the actual detailed work of the design. The concept has to meet the requirements of the brief (which are described in Section 8). This brief will list most of the factors which are going to control the decision, but will still leave the designer with a number of choices. For example, he may have to choose between the use of steel or timber, or perhaps the use of proprietary components. It is important that all these parameters are recited as a preliminary to the calculation, so that subsequent checking or alteration can start with a clear appreciation of the basic parameters. It cannot be over-emphasised that falsework design is just as important as permanent works design, and that corners should not be cut on account of its temporary nature.

The collection of information under the title of the design brief should include as much information about the problem as possible. It will naturally include the drawings of the permanent works; it will draw on any local knowledge of the state of the site in earlier times, perhaps of floods or other natural phenomena to which it is subjected. Where the site is located in an exceptionally windy area steps should be taken to make provision accordingly.

In addition, construction preferences should be included in the brief. The organisation which is going to do the construction will have certain expertise available; it will have tried various methods previously and may have particular equipment readily available. Consideration must be given to the various stages of the construction, the interaction of different parts, and the stability at the various stages. There may be particular requirements for access which need to be considered. It may be intended to move the falsework in a partly erected state from a first position to a later one. Construction may include prestressing, which frequently will cause a major redistribution of the load on the falsework; however, redistribution of load can take place for other reasons as well during subsequent construction.

Another area which requires study and incorporation in the brief is the requirements of neighbours, be they property-owners or statutory authorities. Where falsework is constructed over a river or near a road for example, those authorities will have requirements which must be met. There may be underground services which need to be protected. If there is a river, it may be necessary to protect the falsework against floating debris.

Such information and any other information which could be relevant, should be assembled into the design brief, so that the design process may proceed rationally, without having to backtrack when information comes to light late in the process. It is necessary, both at the start of erection and prior to loading, to check that none of the basic parameters have changed. Should they have done so, the design process would have to be reassessed to see if any changes to the falsework are necessary.

Falsework is very similar to permanent design, but there are factors which are different. It will only be up for a relatively short time, and almost always most or all of the equipment will be subsequently used, in a situation which is frequently not identical. This means that the falsework is likely to be an assembly of components which can be readily dismantled rather than the construction of an indivisible whole. Because it is a collection of small or relatively small pieces, there will be numerous joints. It is important that these are designed so that they will go together correctly and easily so that the structure is not rendered useless by faulty joints. Since falsework is usually removed from below, provision must be made for a suitable stripping method which can take place without difficulty.

There are considerable advantages in designing falsework so that if any one piece fails, the whole structure does not fail. However, it is often due to this duplication of paths that the concept becomes structurally indeterminate. It may even be quite unanalysable; it is frequently better to provide a structure with a simple structural concept which is easy to visualise, so that one can be satisfied that its behaviour will be as predicted.

Standard design or purpose-made

The approach to the design of falsework will take one of two courses. Much falsework is highly repetitive and of very similar application; standard designs are available in the Falsework Code[1] and from some manufacturers of equipment. (See Section 6.) Individual construction organisations also produce their own standard arrangements. In the second course, where these are not available, or some aspect of the falsework diverges from that envisaged in the standard solution, it is necessary to design the falsework from first principles. There will be situations where some tabular information will enable the design to proceed, while it will not be possible to complete the design on a standard solution basis.

Some concepts

There are two main ways of dealing with the problem associated with the common application for falsework, where the load due to the construction of a slab has to be carried directly down to an appropriate foundation. The first is to provide a support with many small capacity verticals, for example with scaffold either tubular or proprietary. Alternatively, the falsework may duplicate the structure itself in having horizontal beam members which are supported in turn on heavier duty vertical supports adjacent to the structural support of the permanent structure. A variant on this approach is to have the horizontal beam members supported above the slab and to hang the formwork and its beams from them. It is likely to be possible to support the beams from the structure itself.

Forces

There are a number of forces which the falsework must resist; it is necessary to consider combinations of these, to make sure all conditions are considered. But to consider all the worst conditions at the same time may be unrealistic. For example, as people will not work in a high gale, the live load of operatives and maximum wind loading would not occur together.

Vertical loads include the load for which the falsework is erected, the self-weight of the falsework itself, as accurate an allowance as possible for the operatives and any equipment which they are using, and any items likely to be stored on the falsework or slab, or construction to be built off it.

Wind load should be the maximum likely to occur in a two-year period for the area, but consideration should also be given to the maximum working wind force corresponding to Beaufort wind scale six which gives a pressure of 200 N/m^2. There will be, however, cases where only forces other than wind need to be considered, for example where the falsework is surrounded by walls.

There is a force due to the erection tolerances. If a vertical member is not truly vertical, a horizontal component is generated. In the case of a single member this may be as high as 3% of the vertical load, though when many members are present, (normally taken as greater than four) a figure of 1% may be taken. This force can act in any direction and is in addition to wind load, not a substitute for it. Should the vertical member be out of plumb by design, the appropriate calculations should be made. In no case should the total horizontal design force be less than 2½% of the vertical load.

Concrete pressures may exert a force on the falsework, but in general it is only dead weight of concrete which needs to be considered. However, formwork must always be designed to restrain the concrete pressure. The forces between the opposing sides are transmitted via the formwork underneath and may impose lateral forces on the falsework. It should be noted that if fresh concrete is to be placed against existing concrete, this arrangement forms part of the formwork system, and the formwork panels must be effectively connected to this pre-existing concrete, otherwise the lateral force generated will push the formwork away. Prestressing can result in significant movement of the concrete (as can thermal effects) in large structures.

Cases to consider

Falsework should be designed so that individual members do not fail. This could lead to progressive failure of the structure and even if it does not, it is likely to lead to local deformation. The structure as a whole should be considered for stability and the possibility of overturning of the falsework should be assessed. The various stages of construction may require separate checks. In order to design, it is necessary to know what tolerances will be used on site. Specific information on this can be found on pages 7, 14 and 33. Only when tolerances which are specific and practically obtainable are adopted, can the design be a reasonable representation of what actually happens on the site.

Design of falsework is still carried out using basic deterministic theory, as the information necessary for a probabilistic or limit-state approach is not yet available. The factor of safety which is normally adopted is two, although this will be increased in adverse circumstances. For steel struts, a reduction is normally made and a figure of 1.7 is used in the strut formula of BS 449[15] (though a higher figure is used for scaffold tube).

Not only should the whole structure be stable, but it is also important that individual components are stable. For example, the head of a column needs to be located in space so that it cannot move. In scaffolding the vertical member will be divided into several parts, each of the points of location constituting a node point which is located in space. Failure to provide this location will invalidate the assumptions of the calculation. It is usual to take a figure of 2.5% of the vertical load as a notional load to ensure stability of node points. Beams may require stabilising to prevent lateral buckling and careful consideration should be given to this point, which is even more important with cantilevers. Steel members of I section may require web stiffeners. It is prudent to install web stiffeners, unless calculation has shown them to be unnecessary. The Falsework Code[1] provides information on calculations. All these stability forces should be carried back to a rigid support point. To tie two points together will avail nothing if they can both fail in the same direction.

Falsework is frequently erected by stacking one component on another. This means that it has no tension capacity and friction has to be relied on in a transverse direction. It will only be in the direct compression direction that the strength is clear-cut. The Falsework Code[1] provides some information on coefficients of friction.

Unless it is completely certain that vibration will not result in movement, an alternative load path should be provided, even though the coefficient of friction is theoretically adequate.

The ultimate support of the falsework is often fundamental in deciding the concept. It is clearly economical to use the existing structure where this is practical, as the cost is already borne elsewhere. But the structure must not be overstressed. There is the possibility that the eccentric loading which almost inevitably will result will be excessive and additionally the cost of providing horizontal beams of adequate strength to transmit the loads of the span to these supports may also prove to be very high. Where the ground is to be used, timbers or concrete should be used to ensure that the load is spread over an adequate area of the ground to prevent local failure. Where there are heavy loads under substantial columns, it may be necessary to construct concrete spread footings or to use piles.

Where heavy beams are in use, the forces created at the hinge point can be important and provision must be made for lateral movement due to deflection under load and thermal movements; rotation which can occur here will also be significant. A low friction hinge detail or rocker should be provided unless it can be shown that the moment so induced will be satisfactorily carried by the supporting tower.

It is attractive to move falsework from one position to another with the minimum of dismantling. Considerable weights can be slid from one position to another if suitable trackways are provided. But it imposes considerable stresses on the assembly and it is most important that the design is examined for this mode of use. The falsework should be carefully examined in its new position to ensure that it is not distorted and is in exactly the right place.

The avoidance of eccentricity is always important.

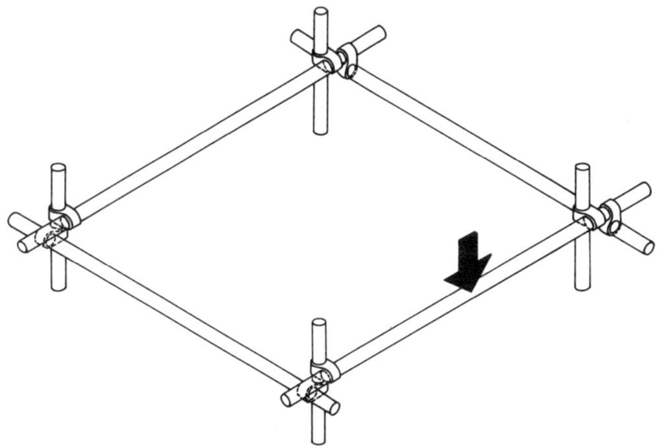

Fig. 16. Diagram of scaffolding showing eccentric loading

Beams should be centralised over vertical support members. Joints can be made by butting in a forkhead, but a high standard of accuracy is needed. It is better to lap them, though this may create a difficulty with intermediate supports. All overhangs should be kept short. Beams are likely to be laterally unstable if their height–to–base ratio exceeds two to one. Additional support is required. If they have to be canted, lateral stability demands particularly careful attention.

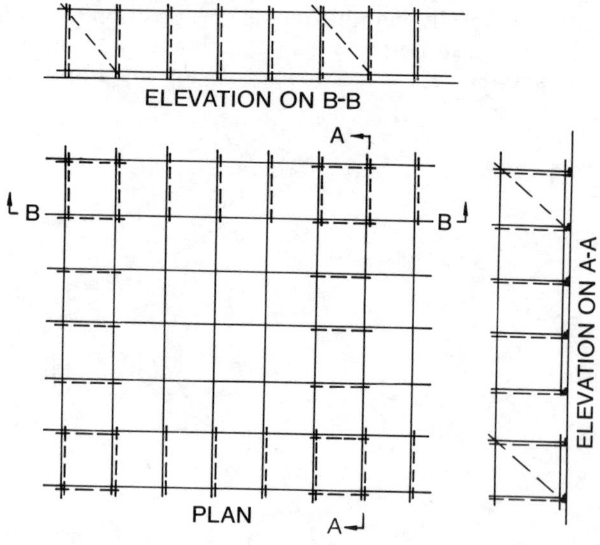

Fig. 17. Arrangement of bracing in a scaffold birdcage (broken line indicates bracing)

Where beams cross each other in different planes, it may be practical to use tapered packs to provide the load path. Bearing stress requires careful consideration. Waney timber not only produces higher bearing stresses but an eccentric load as well, and should not be used in bearing.

Use of scaffold

Because couplers bring the load on eccentrically, there is potentially a significant reduction in the capacity of a standard. This can be minimised if there is support at the other end, which will counteract the cantilever effect. This is shown in Fig. 16. Normally, however, loads should not be brought on to a scaffold structure via the horizontal tubular members. It is much more efficient to bring the load on concentrically through the end of the tube. This is done by an adjustable post-head or forkhead at the top, and where necessary an adjustable base at the foot. The effective length of tubes in a scaffold structure should never be taken as less than one. Where a tube projects beyond the structure, the section between the last two node points is taken as L and the cantilever as mL, and the effective length will then be $L + 2mL$. This shows that top and bottom bays and cantilevers are generally the most critical parts of a scaffold. While this is normally a problem at the top, the same situation occurs at the bottom, if the level of the first or second level of lacing is not connected to ground—an arrangement which good practice requires for other reasons.

In a typical birdcage scaffold (Fig. 17), stability must be provided. This may be achieved by locating the several

Fig. 18. Use of prefabricated scaffold to support a bridge deck

Fig. 19. Typical construction based on standard solutions

levels of lacing against a surrounding structure or by attaching them to columns. But often it is necessary to resort to bracing tubes. These should be fixed in elevation ideally at 45° but normally between 30° and 60°, at a rate of at least one per five standards.

If in doubt, calculate. Where there is a substantial strong point, like a bridge column, some bracing can be omitted from lines of standards which are effectively anchored to them. But this lateral restraint cannot be assumed to provide support more than six standards away from the support.

Braces should be attached at node points and should end as near the top and bottom of the scaffold as possible. At the bottom the horizontal components of load must be resisted by the ground or some other substantial object. It is good practice to use only right-angle couplers for connecting bracing tubes; this requires careful consideration when setting out the scaffold. If it is not done accurately, braces cannot be fixed without being bent, or using swivel couplers. Joints in braces should be made with sleeve couplers because they may go into tension. Ideally, joints in braces and all other tubes should be at the quarter points, but in any case adjacent members should not have joints in the same bay. Joint pins may be used in vertical members when there is no tension, otherwise sleeve couplers are essential.

Prefabricated scaffolds (Fig. 18)
The philosophy for tubular scaffold applies equally to prefabricated scaffold. However, most prefabricated scaffolds have in-line joints which eliminate the eccentricity. The joints are normally stiffer than ordinary couplers, but it is still necessary to provide bracing. The bracing is so much more effective than the joints that they add little in a braced bay. The stiffness is useful during erection and is the main reason why prefabricated scaffold provides good vertical capacity. Scaffold made with frames is largely self-bracing, but for all prefabricated scaffold it is essential to follow the design rules given by the supplier.

Effects of inaccuracy on the design
The design concepts in the Falsework Code[1] assume that there will be minimal deviations from the ideal. There will be some eccentricities and some members may not be truly vertical or horizontal. Provided erection is kept within the specified tolerances, the design can be carried out without further analysis of these points.

Tubular scaffold
Verticals should be plumb within 15 mm over 2 m of height, subject to a maximum displacement from the vertical of 25 mm, and eccentricity of load at top or bottom should not exceed this. The centre-lines of tubes at a node point should be as close as possible and should never be more than 150 mm apart. Sole plates should be level within 1 in 40.

Props
See 'Manufactured equipment', pages 12 and 14 for the acceptable tolerances.

Timber
See 'Timber', pages 5 and 6. Some American timber has a greater deviation from the nominal size, and the design should take this into account.

Structural steelwork
Verticality. For vertical members the accuracy is assessed over any length L.

$$\text{Where } L < 1450 \text{ mm}, \Delta < 5 \text{ mm}$$
$$\text{Where } 1450 \text{ mm} < L < 7143 \text{ mm}, \Delta < 0.0035L \text{ mm}$$
$$\text{Where } L > 7143 \text{ mm}, \Delta < 25 \text{ mm}$$

where L is the overall length in mm and Δ the amount it is out of plumb. Eccentricity at a bearing should not exceed 5 mm.

Calculation

Calculation for steelwork follows conventional structural design, but reference can be made to *Scaffold falsework design.*[16]

6. STANDARD SOLUTIONS

It is always possible to design falsework from first principles, but many of the problems are repetitive and to do the same calculations over and over again is unproductive. Standard solutions are applied in the context of frequently used equipment, both traditional and proprietary. Information for common slab thicknesses and beams is given for example in the Falsework Code,[1] using props or scaffold, and timber runners and bearers. Provided the associated rules for use are followed, technically satisfactory falsework will result. This can be a very convenient way of producing falsework, where problems are simple and design facilities are not available, but it may not produce the most economical solution. Fig. 19 shows a typical arrangement for which a standard solution would be appropriate.

Such solutions should be presented with precise limitations. The minimum grade of materials and the particular equipment to be used must be clearly specified; the accuracy of construction will be set down. Normally both the state of repair of the components and the level of accuracy required will be that given in the Falsework Code.[1] (See also Sections 3, 5 and 7 of this Guide.) Such standard solutions should be carefully examined to make sure that all the assumptions adopted by their designer have been included in the rules for use. Sketch layouts should be produced, so that no one is in any doubt as to what is to be erected. Provided its calculation is in line with the Falsework Code,[1] such a standard design will be in accordance with the Code when used.

If the circumstances of the job are in any way outside the limitations of the particular standard solution, then a full design will be needed.

7. QUALITY AND CHECKING THE FALSEWORK

If the falsework is erected with substandard or incorrect components, it may be dangerous. If it is put up carelessly or deviates from the designer's intentions, it will not behave as he anticipated. The standard solutions given in the Falsework Code[1] are supported by detailed requirements for the quality of the components and material to be used and the accuracy with which they are to be erected. (See also Section 5.) These represent a standard which it is practical to achieve under site conditions. A similar level will apply to all other falsework unless the designer has chosen to adopt a different basis. This section is devoted to checking the falsework.

Brief

It is important to check that the conditions of use of the

falsework are those envisaged at the design stage and that nothing has changed for the worse. Are foundations as secure as expected? Are there any unforeseen loads? Have concreting methods changed? Is the order of loading the same? There may be a considerable period between design and execution.

Drawings

The erected falsework should be checked against the final design given in the latest edition of the drawing and any supporting paperwork such as a method statement. It is not possible to check falsework which has no drawings; calculations will be necessary.

Correct items

It is important that all materials or components used are those intended, both in quantity and quality. For example, it is often not easy to be sure that the quality of timber is good enough; steel sections come in different weights and grades which are superficially indistinguishable. Proprietary components are frequently available in models which are similar but not of equal capacity and there are very similar sized metric and imperial versions. In addition, there must be no undue deterioration or damage. Bent or corroded steelwork will be weaker than when new as will material with holes or notches. Damaged components could be the weak link in the falsework.

Identification of materials and components

While such material and equipment will rapidly become familiar, it is desirable to mark anything where doubt can arise, and to obtain identification literature from suppliers. It is remarkable how often similar but not identical components get mixed up.

Steel. The thickness and thus the weight of standard rolled sections of steel can be established by careful measurement and reference to tables. But the quality of steel can only be established by a laboratory. It is essential to mark steelwork as permanently as possible in accordance with the delivery note when it arrives.

Timber. Size of timber can be checked, but species and quality (or grading) are difficult for the non-expert. Stress-graded timber will be already marked. It is also necessary to mark other timber which is not of the lowest class used on site when it arrives.

Components. In a few cases, components will be marked; otherwise it is necessary to compare them with the manufacturer's identification literature. In case of doubt, the supplier's representative should be contacted.

Accuracy

The way the various parts are assembled is usually critical. Relatively small inaccuracies in setting out can result in major changes in the way the falsework carries the loads. It goes without saying that the correct number of components should be put near enough in the right places. But how near is near? This is discussed in Section 5. The minimising of eccentricities is of particular importance. Ideally there should be none, but even where the layout precludes their elimination, the lack of straightness of components, and the extended time needed to adjust everything to a tolerance of ± 1 mm result in the acceptance of a lower standard. The more practical level actually adopted is taken account of at the design stage, but it must be regarded as an absolute limit and not just a target during erection.

Dimensions

It is important that the falsework structure as a whole is in the right place. The load must come on it correctly, so that the individual members each carry their intended share. The distances between components must be right.

Falsework will deflect under load and the design should show if for example beams require any setting up to counteract this effect. A stack of components will 'take up' under load at each joint and allowance should be made for the loss of height.

Access

Working platforms will normally be shown on the drawings, but must be reconsidered at the checking stage. It is then easier to see if adequate provision has been made and a careful examination should be made to ensure there is proper compliance with the Working Places Regulations.[17] At the same time a look should be taken to ensure there are no traps or hazards. BS 5973,[18] the Code on scaffolding, provides information on various types of scaffold.

Checking and checklists (see Appendix, p. 41)

Naturally, falsework should be put up using the correct equipment in adequate condition in the correct place. However, it is essential that this is checked. It is difficult to remember all the various facets. Thus it is useful for the checker to have a list of points. For simpler applications, a standard list or lists may suffice but it is often more satisfactory to use these as a basis for a purpose-written list. It will be necessary to establish lists for proprietary equipment or obtain them from the supplier.

The list opposite summarises the points to look for. The detail has already been discussed and the necessary information on what is acceptable will be found in earlier sections or in the case of scaffold components can be obtained from *Access scaffolding*.[8] If a designer has adopted different standards, he must provide the necessary information to the people directly responsible for the erection. For proprietary equipment, the supplier should be contacted. The lists in the Appendix are divided into assessing the material and equipment, and erection. For particular jobs, it will be helpful to draw up checklists which are directly relevant. These can be supported by diagrams or pictures showing what to look for. Examples are given in *Construction safety*.[19]

Checking is likely to be appropriate at intermediate stages in the erection as well as when it is ready to load. In this way abortive work can be reduced as it is often difficult to assess foundations later.

If a check is not 100% successful, it will be necessary to check again after the faults have been put right; thus a checklist with three columns will be appropriate: check, faults found, and final check.

8. CONTROL

Falsework must be designed, erected, loaded and finally dismantled when its purpose is completed. Experience shows that this simple sequence can go wrong very easily in one way or another. Thus the management or control of these activities is of prime importance. Many different people may be involved and it is essential that collaboration is complete. It should be understood that management encompasses everyone from project manager to ganger or charge-hand.

Because of public concern about collapses, the government set up a committee to make proposals for improvements. This resulted in the Bragg Report[20] on which some of the information in this Guide is based.

Action

Design

In order to produce a design, all the parameters which can affect the outcome must be assembled, ideally into a single brief. They will include

(*a*) details of the structure to be supported, and any ways in which its behaviour may be at all unusual
(*b*) programme limitations
(*c*) details of any equipment or material which it is hoped to use and availability of others
(*d*) information about the ground or structures on which it may have to be founded
(*e*) facilities for or limitations caused by the public, neighbours, or statutory undertakers
(*f*) risks from water, flooding or erosion

The completed design should be independently checked for concept and structural adequacy.

A man who has helped or guided the designer cannot be regarded as independent; the larger and more complex the falsework, the more independent the checker should be. Ideally he should not examine the calculations, but do his own independent analysis of the proposed falsework structure. When detailed checking of the design has to be carried out, the calculation method adopted must be assessed as well as the accuracy of the arithmetic.

Erection

The man in charge should receive the design with all its associated details and study it until he is familiar with it.

Materials must be ordered. When they arrive they must be examined to see that they are those required and are in appropriate condition. During erection, each piece will receive an inspection, albeit cursory, from the operative who handles it and management must make clear what is and what is not acceptable.

The setting out of the falsework must be carefully done to ensure that it will be in the right place for the load to come on as intended.

The foundations must be prepared and if the ground is not suitable for the pressures adopted in the design, some redesign will be needed. This must be referred back to the designer.

Where material is being reused, it should be examined

as it is being re-erected to check it is still in good enough condition. Care must be taken to ensure that the falsework is in accordance with the design and within tolerance. The accuracy of erection must be constantly under review, so that expensive rebuilding is not needed, or worse still, inaccurate erection be accepted due to the pressures of cost and the need for progress. When the manager requests inspection of the falsework, he should do so with the confidence that he will not be embarrassed by having to alter it.

Inspection must be carried out by someone who has not been directly involved in the erection. To have been party to the faults which he identifies would put him in a totally unacceptable position. Except in the case of the smallest falsework, it is very sensible to check at more than one stage. Early on is the time to check the basic setting out and the foundation—it may be too late afterwards. The final check can only take place when all the promised extra bracing has actually been completed. No check is complete until the checker is satisfied the faults have actually been remedied.

Loading

Before loading, the Falsework Co-ordinator (see opposite) should sign a permit to load. He will do this only after a satisfactory inspection has been completed, and the circumstances compared with those anticipated in the brief. The permit should make quite clear the extent of the approval in terms of area and the load to be applied. It should be of limited validity—a day or two—to reduce the risk of later modifications.

Dismantling

The falsework must not be dismantled before the permanent structure is able to support itself. The exact requirement may be set down in the specification, or the permanent works designer may issue instructions on the point. Alternatively it may be necessary to calculate or assess when dismantling is possible. The Co-ordinator should issue a permit to dismantle.

Any material which obviously cannot be reused should be scrapped.

Falsework Co-ordinator

Managers should carry out all the actions described in this Section without further check. But experience shows that having someone else with the duty to see that the checks are all done and done properly is highly desirable; such a man is called a Falsework Co-ordinator. The one direct duty he has is signing permits to load or dismantle, otherwise he has only to ensure that others do their job. Ideally he should have no other tasks but in practice he may well undertake some of the actions.

Subcontracting

Where one organisation is in charge of the falsework, and everyone is a direct employee, it is not difficult to allocate and control all the various activities, reducing to a minimum the risk of misunderstanding or something not being done because it falls between two people. Where there are two or more separate organisations, the problems are greater; the need to establish exactly who is doing what is much more urgent.

The role of the permanent works designer

Under the typical British contract system, the contractor undertakes to erect the permanent structure, in accordance with the drawings and specification including any necessary falsework. The client has a duty to supply all necessary information to enable this to be done, which he does by asking the permanent works designer to provide the information (as his agent). The designer, probably as Resident Engineer, will also normally be expected to protect his client's interests (time, reputation) by forestalling any difficulties. Thus, he may decide to study (but not approve) the contractor's intended approach. If he intervenes because he is not convinced the proposed method is satisfactory, he may involve his client in greater costs. If he fails to exercise his skill in preventing a disaster, he may find his organisation legally responsible for a share of the consequential costs. Thus it is important that he decides which falsework to examine and then does so to the best of his ability.

In a similar manner, if he is on site, he must keep an eye on what is happening and decide if and when he should make a detailed examination of what is being erected.

Personal responsibility

The common-law responsibility of duty to one's neighbour requires anyone with appropriate skill and authority to exercise his skill and expertise to protect the neighbour, and this applies to all persons with responsibility on site, no matter who employs them. Failure to perform up to the limit of a man's ability could result in legal consequences.

ACKNOWLEDGEMENTS

The Author is grateful for the assistance of many friends in the industry, without whom this Guide would not have been written.

In particular, the following companies and organizations have granted permission for the reproduction of photographs or diagrams:

Acrow (Engineers) Limited (Fig. 7)
Construction Industry Research & Development Association (Fig. 8[11])
Institution of Civil Engineers (Fig. 1)
John Laing Design Associates Limited (Figs 3, 4, 15, 16, 17 and 19)
Kwikform Limited (Figs 9 and 10)
Mabey Bridge Company Limited (Fig. 13)
McGraw-Hill Book (UK) Ltd (Fig. 5)
G K N Mills Building Services Limited (Fig. 12)
Rapid Metal Developments Limited (Figs 2 and 14)
Scaffolding (Great Britain) Limited (Figs 6 and 18)
Thomas Storey Limited (Fig. 11)

REFERENCES

1. BRITISH STANDARDS INSTITUTION. *Code of practice for falsework*. British Standards Institution, London, 1982, BS 5975.
2. BRITISH STANDARDS INSTITUTION. *Code of*

practice for the structural use of timber. Part 2: permissible stress design, materials and workmanship. British Standards Institution, London, BS 5268. To be published.
3. BRITISH STANDARDS INSTITUTION. *The structural use of timber. Part 2: metric units*. British Standards Institution, London, 1971, CP 112.
4. BRITISH STANDARDS INSTITUTION. *Timber grades for structural use*. British Standards Institution, London, 1976, BS 4978.
5. TIMBER RESEARCH AND DEVELOPMENT ASSOCIATION. Simplified rules for the inspection of second hand timber for load bearing use. TRADA, High Wycombe, 1981.
6. BRITISH STANDARDS INSTITUTION. *Metal arc welding of carbon and carbon manganese steels*. British Standards Institution, London, 1974, BS 5135.
7. BRITISH STANDARDS INSTITUTION. *Metal scaffolding. Part 1: specification for tubes for use in scaffolding*. British Standards Institution, London, 1982, BS 1139.
8. WILSHERE C.J. *Access scaffolding*. Thomas Telford Ltd, London, 1982, ICE Works Construction Guide.
9. BRITISH STANDARDS INSTITUTION. *Metal scaffolding. Part 2: specification for couplers and fittings for use in tubular scaffolding*. British Standards Institution, London, 1982, BS 1139.
10. BRITISH STANDARDS INSTITUTION. *Metal props and struts*. British Standards Institution, London, 1982, BS 4074.
11. BIRCH N. *et al.*. *Safe working loads for adjustable props: the influence of prop condition and site workmanship*. Construction Industry Research and Information Association, London, 1977, TN 79.
12. BRITISH STANDARDS INSTITUTION. *Methods of test for falsework equipment. Part 2*. British Standards Institution, London. To be published.
13. BRITISH STANDARDS INSTITUTION. *Methods test for falsework equipment. Prefabricated heavy duty support systems—methods of measurement and assessment of performance*. British Standards Institution, London, DD89. To be published.
14. HATHRELL J. (Ed.) *Bailey and Uniflote handbook*. Acrow Press, London, 1975.
15. BRITISH STANDARDS INSTITUTION. *The use of structural steel in building. Part 2: metric units*. British Standards Institution, London, 1970, BS 449.
16. GRANT M. *Scaffold falsework design to BS 5974*. Eyre and Spottiswoode Publications Ltd, Andover, 1982, Viewpoint Publication.
17. The Construction (Working Places) Regulations 1966. No. 94. HMSO, London, 1966.
18. BRITISH STANDARDS INSTITUTION. *Code of practice for access and working scaffolds and special scaffold structures in steel*. British Standards Institution, London, 1981, BS 5973.
19. BUILDING ADVISORY SERVICE. *Construction safety*. BAS Management Services for the National Federation of Building Trades Employers, London, 1982. Section 21.
20. BRAGG S.L. (Chairman). Final Report of the Advisory Committee on Falsework. HMSO, London, 1975.

APPENDIX: FALSEWORK CHECKLIST

1. General	Check	Faults to be put right	Final check
Compliance with brief:			
Is the brief still correct? . . .			
Compliance with drawing(s):			
position			
dimensions			
tolerances			
correct components			
condition of components . . .			
Environment:			
protection against flood . . .			
protection against traffic . . .			
adjacent excavation			
other interference			
Access:			
working space			
decking			
guard rails			
toe boards			
ladders			

continued overleaf

1. General

continued

Stability:

node points . . . .

whole scaffold . . . .

across adjustable components

tall beams . . . .

jointing of beams . . . .

overhang . . . .

Movement:

allowance for take-up . . . . . .

allowance for deflection . . .

Check	Faults to be put right	Final check

2. Foundations

	Check	Faults to be put right	Final check
Basic support:			
structure below			
the ground			
risk of differential settlement			
frozen ground			
swallow holes			
Sole plate (mud cill):			
condition			
position			
adequate area to receive load			
stiffness			
bedding			
protection against deterioration			
Sloping ground:			
specialist advice			
proper load path			

3. Timber

Correct material:

size . . . .

species . . . .

grade . . . .

condition . . . .

damage . . . .

Joints:

position . . . .

cantilevers . . . .

continuity . . . .

Bearing above and below:

adequate area . . . .

eccentricity . . . .

tapered packs . . . .

wane . . . .

Lateral stability:

narrow timber . . . .

slopes . . . .

Check	Faults to be put right	Final check

4. Steelwork

	Check	Faults to be put right	Final check
Correct material:			
size			
weight of section			
grade of steel			
condition			
damage			
Position:			
location			
tolerance			
Details:			
accommodation for movement			
web stiffeners			

5. Tube and Fitting Falsework	Check	Faults to be put right	Final check
General:			
adequate for application ...			
in accordance with design ...			
Correct material—tube:			
type			
dimensions			
condition			
straightness			
Correct material—fittings:			
type			
parts in working order ...			
Correct components:			
correctly tightened			
no tube mixing			
base plates			
tube			
couplers			
sleeve couplers			
forkheads			

continued on facing page

5. Tube and Fitting Falsework	Check	Faults to be put right	Final check
Dimensions:			
position			
horizontal			
vertical			
adjustable extensions			
Tolerances:			
vertical			
level			
node points			
adjustable extensions			
Bracing:			
quantity			
dimensions			
tolerances			
forkheads			
foot ties			
Tube joints:			
horizontals			
standards			
bracing			

6. Telescopic Props

Correct equipment:

pin . . . .

unbent . . . .

uncorroded . . . .

Erection:

full contact with base . . . . . .

plumb within 1.5° of vertical

loaded within 25 mm of centre (or concentric) . . . . . .

laced if taller than 2.75 m . . .

Check	Faults to be put right	Final check

7. Telescopic Centres

Correct material:

- unbent … … … … …
- uncorroded … … … …
- welds uncracked … … …

Erection:

- spacing and span … … …
- minimum overlap maintained
- locking device tightened …
- end lugs fully supported …
- lugs protected from concrete
- supports stable against overturning … … … …
- camber and deflection allowed for … … … …
- intermediate propping only to suppliers rules … … …

Check	Faults to be put right	Final check

8. Aluminium components

Specified equipment: ... ...

unbent ... ...

uncorroded ... ...

Erection:

position ... ...

location ... ...

tolerance ... ...

Check	Faults to be put right	Final check

9. Proprietary Equipment	Check	Faults to be put right	Final check
Correct material:			
components			
Condition:			
unbent			
uncorroded			
unmodified			
welds uncracked			
Dimensions:			
position			
horizontal			
vertical			
adjustable extensions			
Tolerances:			
vertical			
level			
adjustable extensions			
Bracing:			
quantity			
precise location			
Access:			
assembly			
use			
dismantling			

ALUMNI/LYDON LIBRARY UNIVERSITY
OF LOWELL